ISBN 978-1-334-18292-1
PIBN 10687270

This book is a reproduction of an important historical work. Forgotten Books uses
state-of-the-art technology to digitally reconstruct the work, preserving the original format
whilst repairing imperfections present in the aged copy. In rare cases, an imperfection in
the original, such as a blemish or missing page, may be replicated in our edition. We do,
however, repair the vast majority of imperfections successfully; any imperfections that
remain are intentionally left to preserve the state of such historical works.

MANUFACTURE AND USES
OF STEEL PIPES

A THESIS

presented by

JAMES S. HARVEY, JR.

to the

President and Faculty

of

ARMOUR INSTITUTE OF TECHNOLOGY

for the degree of

Mechanical Engineer

May Twentieth, Nineteen Hundred Fourteen
CHICAGO, ILLINOIS

F. H. Gebhard

H. M. Raymond

L. C. Monin

Acknowledgements.

The writer is greatly indebted to Miss Josephine McAndrew and Mr. J. Hall Taylor for very material assistance in compiling the article and wishes to thank them for their kindness.

He also wishes to express his thanks and appreciation to the following firms for their kindness in rendering such hearty assistance in the form of data and illustrations:

The American Spiral Pipe Works, Chicago, Ill.

The National Tube Company, Pittsburg, Penn.

The Mannesmann Tube Company, Montreal, Que. Can.

Drummond, McCall & Company, Montreal, Que. Can.

Stewarts and Lloyds, Ltd., England.

The Riter-Conley Constr. Co., Pittsburg, Penn.

East Jersey Pipe Co., Paterson, N. J.

James S. Harry Jr
May 20 - 1914.

Table of Contents.

List of Illustrations.

Fig. 1. Hydro-electric line of Homestake Mining Co.

The necessity of conveying water, gas, oil
and other materials from one place to another with
greater facility was the cause for the experiment-
al and research work, from which developed the
present systems of piping.   In the early ages the
particular need in this direction was for some way
of conveying a supply of water to the towns and
villages, especially in regions where wells were
not sufficient.   This was taken care of by stone
and cement aquaducts.Parts of these were made in
open channels and other portions were constructed
underground.   Where it was necessary to cross a
gorge or a valley, a closed conduit was sometimes
used, but usually an arched bridge of massonry was
erected with channels at the top for carrying the
water.

There were twelve aquaducts which furnished
the water supply to Rome from the various springs
surrounding the city.   Two of these from the
Sabine Hills extend a distance of 62 miles while
the others vary from 6 to 45 miles in length.   The
durability and the archetectural beauty of their
arched bridges attract much notice and the entire
systems with the curves and bends necessary to con-
form to the irregularities of the country are
wonderful examples of what may be accomplished
even under adverse circumstances.

Lead pipes were also used, but not to a very
great extent on account of the limiting size, for
the material could not withstand high pressures.
The Romans also recognized the fact that the lead
had a tendancy to poison the water.   Later on
wood pipes came into use and even in the present
day are used to a very great extent in some sections of the country.   The style of construction
of course is very different.   Originally they
were made from large trees, so that straight pipes
from 12 feet to 20 feet long and from 1 1/2 inches
to 8 inches bore could be gotten.   The bore was
usually equal to one-third of the diameter of the
tree.   It is a well known fact that common pump
logs with bark and sap on, have been used successfully for water mains for several hundred years.
It is recorded that over 400 miles of elm pump logs

Fig. 2-Hydro-Elec. Power Line in Nor-
way made of Männesmann Tubes.

were laid in London, England in 1613. Some of
these pipes were taken up in good condition in
1862 after having been down 249 years. Of course
this type of construction could not be used for
pipes of large diameters. Wooden staves are then
used and the pipes are built up - usually being
reinforced by iron bands at frequent intervals.

The first mention of cast iron pipes was in
the description of an elaborate system at Marti,
near Paris, France, which was installed in 1682.
Water was drawn from the river Seine by short suc-
tion-tubes and forced through cast iron pipes up
the hill. The height of 533 feet was made in
steps by means of a series of reservoirs and pumps.
This system became known as the "Monument of
Ignorance", for it was soon seen that the cast
iron pipes were able to withstand considerable
pressures. Cast iron pipes are used a very great
deal in this present day, especially for water and
gas lines, but on account of their great weight
are very difficult to handle.

In 1824 James Russell of England, succeeded
in the manufacture of tubes made of sheet-iron
strips welded together at the seams. This has
been developed into the lap and butt welded tubes
of today. About the year 1853 wrought-iron pipes
with longitudinal straight riveted seams were used
on the Pacific Coast. This was the usual method
at that time and even more recently; many towns
are still supplied with water under considerable
head through pipes of this kind, which have been
in service more than 30 years.

The first occasion when steel pipes came into
use was in the year 1881 when the Kimberly Water
Works had to buy some 1500 tons of water mains.
There being no railway in at that time and a
necessary haulage of about 500 miles by ox-team,
the cost of the freight alone excelled $200.00 per
ton. It was then apparent that cast iron pipes
were out of the question. Thomas Piggott of
Birmingham offered to furnish welded steel tubes
14 inches in diameter and 1/4 inch thick and 18
inches in diameter and 5/16 inch thick. His pro-
posal was accepted and the line was installed in
that way. Since then in South Africa all steel

water mains have been installed in Pretoria,
Durban, Klerksdorp, Buluwayo, and in the Johanns-
burg Water Works.

11 Foot Riveted Steel Pipe
Torresdale Filtration Plant, Philadelphia, March, 1908

Fig. 3.

Manufacture.

Steel tubes which are manufactured in a
great variety of ways may be grouped into
three general classifications under the heads
of: Welded Tubes; Seamless Tubes and Fabricated Tubes.   These may still be further divided
as follows:

1st. Welded Tubes	1. Butt Welded
	2. Lap Welded.

2nd. Seamless Tubes	1. Punching of solid billets by power presses.
	2. Plate cupping process.
	3. Cast hollow-billet process
	4. Piercing process for round solid billet.

3rd. Fabricated Tubes.	1. Steel plate by forming and riveting seam.
	A. Straight Riveted.
	B. Spiral Riveted.
	2. Steel plate by press- ing and locking seam.
	A. Lock Bar.
	B. Straight Lock Seam
	C. Spiral Lock Seam.

Welded Tubes.

Manufacture of Butt Welded Steel Tubes.

In the manufacture of butt welded steel tubes the material which is used is a low carbon Bessemer or Basic Open-hearth Steel, which is made into plate, termed skelp. These plates are obtained from the mill, rolled to the necessary gauge and width, and cut to length for the particular pipe which is to be manufactured. The low percentage of carbon for welded work is a necessity in order to obtain a good weld at the seam. Experience has shown that certain limits of the composition are necessary.

The following table shows the average chemical and physical properties:

	Chemical analysis				Physical pulling tests.			
	Carbon	Manganese	Su phur	Phosphorus	Elastic Limit	Tensile Strength	Elongation in 8 inches	Reduction in area.
	%	%	%	%	Lbs.	Lbs.	%	%
Bessemer Pipe Steel	-.07	.30	.045	.100	36000	58000	22	55
Open-hearth Pipe Steel	-.09	.40	.035	.025	33000	53000	25	60

For screw joint pipe Bessemer Steel is pre-ferred to Open-hearth Steel, owing to the difficul-ty of cutting perfect threads when the latter is used. When the ends of the pipes are to be flang-ed, Open-hearth steel is preferred. When neither threading nor flanging is required, steel made by either process will answer equally well.

The skelp is first scarfed or beveled slight-ly on the edges so that the edges of the plate when

Fig.4. Drawing Butt—Weld Pipe

formed will meet squarely together.   It is
then heated uniformally to a welding temperature.
When properly heated, the steel strips are gripped
at the ends by a pair of tongs and drawn from the
furnace directly through funnel shaped dies.   The
inner surfaces of these dies are curved so that the
plate is gradually formed into the shape of a tube
and the funnel shape, forces the edges squarely to
gether and completes the weld.   For some sizes of
pipe two sets of dies are used consecutively at one
heat, one being just
behind the other and
the second one being
of a slightly smal-
ler diameter than the
first.

The pipe then
passes through a pair
of sizing rolls plac-
ed one above the
other and operated by
power.   Each of these
rolls has a semi-
circular groove thus
forming a circular
pass corresponding to
the size of the pipe
being made.   Any ir-
regularities in the
pipe are corrected in
these rolls and the
exact outside diamet-
er is obtained.

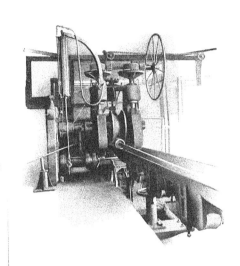

Fig. 5 - Welding Rolls.

Finally the tube is run through
the straightening or cross rolls which consist of
a pair of rolls set with their axes at an angle
with each other.   These rolls have their surfaces
curved in such a way that the tube is in contact
with each for the whole length of the roll.   It
is rapidly rotated then and passed forward at the
same time while the rolls revolve.   The last
operation through the cross rolls makes the pipe
straight and also gives it a clean finish.   The
pipe is then slowly rolled up an inclined cooling
table so that the metal will cool off slowly and
uniformally and maintain its roundness.   Butt
welded steel tubes are manufactured in all sizes
from 1/4" to 3 1/2" in diameter, larger than this
the lap welded process is used.

The Manufacture of Lap Welded Steel Tubes .

Lap welded steel tubes, like the butt welded tubes are manufactured of a steel of a low carbon content. The lap welding process consists of two operations, namely: the bending of the steel sheet and the welding of the seam. There are several methods of performing the weld.

In the first method, the plate which has been received from the mill rolled to the necessary width and gauge, is brought to a red heat in a suitable furnace and is then passed through a set of rolls which bevel the edges so as to give a smooth, even seam when the plate is over lapped and welded. It then passes immediately to the bending rolls where it is formed into a cylinder-ical shape with the two edges over lapping. In this form it is heated in another furnace and when brought to the welding temperature it is pushed out of the furnace onto the welding rolls. Each of these rolls has a semi-circular groove which forms a circular pass, corresponding to the size of the pipe to be made.

A cast iron ball or mandril which is shaped like a projectile, is held in position between the rolls by a heavy bar, and serves to support the pipe as it slides over it and passes through the rolls. In this operation the lap joint is solidly compressed together by the action of the rolls on the mandril. The lap is welded solidly together and reduced to the same thickness as the rest of the pipe.

The pipe is then run through the sizing and cross rolls similar to those used in the butt weld process which straighten it and give it the correct outside diameter and finish. It is also rolled and cooled in the same manner. Pipe is made by this process from 2 inches to 30 inches in diameter and 1 1/8inches in thickness. This method is the one which is used in the United States and England for lap welded tubes.

Germany, which is further advanced in the in-dustry than any other nation, has another method which is also used to some extent in this country.

Fig. 6.   66" I.D. x 1" thick Lap Welded Steel Tubes

Fig. 7. Tube with welded flanges and follower rings.

By this method lap welded pipes may be made to practically any necessary diameter and thickness up to 1 1/8 inches. Thicker than this, it is difficult to make a perfect weld as the heat does not penetrate the sheet.

The first operation is the same in regard to rolling the sheet into a cylinderical shape with the edges over lapping. Then instead of heating the whole tube, the pipe is heated locally along the seam by means of a gas flame, and when the right temperature is reached, the weld is performed by hammering. This makes a very fine weld and is one of the points of advantage which this method has over the first as the hammering condenses the structure of the steel and brings the molecules more closely in contact with each other. After welding, the tube is placed in the furnace and heated to a bright orange color in daylight (1750 deg. F.), which thoroughly anneals it and removes any strains set up in the metal by welding. It is then rerolled to an accurate diameter and straightness.

In pipe over 44 inches diameter and 20 feet long, it is necessary to have two longitudinal welds as the largest sheet rolled in this country is 300 inches by 140 inches by 1 inch thick.

Some German concerns in welding tubes over one inch in thickness of wall do so by scarfing the edges of the plate, fitting them together and welding in, a separate square bar. This joint requires very careful and competent labor, and as it leaves two seams side by side, it is considered inferior to the lap weld, especially to the long joint lap weld obtained by extended scarfing of the corners of the lap.

The Actien - Gesellschaft Ferrum of Zawodzie bie Kattowitz, Germany, has manufactured some steel pipe for exceptionally heavy pressures in connection with high-head water power developments. In order to provide stronger pipe than that given by lap welding of plate of the necessary thickness, the construction that they employ is that of banding the pipe with steel rings or banks rolled out of a solid billet and shrunk on to the plain pipe. In making a pipe of strength equivilent to a thickness of 2 3/16 inches, a seamless rolled pipe core of

Fig.8. Large Hot-Drawn Seamless Tubing.

1 inch thickness is used and the rest of the
strength is obtained by putting on bands. In
order to do this economically, the size and spac-
ing of the bands are calculated so that the internal pressure will strain the material between
the bands and the bands themselves to the elastic
limit at the same time. It has seemed desirable
to use a few bands of large cross section rather
than many smaller bands.

Professor R. T. Stewart of the University of
Pittsburg, found from numerous tests that butt-
welded wrought iron pipe is 70 percent as strong
as similar butt-welded steel pipe, and that lap
welded wrought iron pipe is 57 percent as strong
as similar lap welded steel pipe. In steel the
butt weld averages 73 percent of the tensile
strength, and the lap weld 92 percent of the tensile strength of the material.

Fig. 9 - Section of 66" Exhaust Steam Header.

14

- Seamless Tubes -

In the manufacture of seamless tubes there are four processes which are used, although some of them not to a very great extent.

First is the method of punching the solid billet by means of a hydraulic or power press . This is followed up by heating and rolling or by hot drawing. Either of these operations reduce its wall thickness and elongate it to its necessary length. In the second process, the billet is cast hollow and this hollow cylindrical blank is elongated and reduced in wall thickness either by the hot drawing process as before, or by rolling over a plug or mandrill.

The third is a plate cupping process which is used in making tubes from 5 to 20 inches in diameter. In this process a square steel plate is used which varies in size from 2 to 6 feet across and 3/8 inch to 3

Fig. 10 a - Hot-Drawing Bench.

inches in thickness according to the size tube to be made. The corners are first sheared off to produce a circular disk, which is heated to redness and then placed on the anvil of a large hydraulic press. This punches it into a rough shallow cup similar to those shown in Fig. 10 c. The cup is again heated and punched through a smaller die to deepen it and also reduce its diameter. Possibly several such operations occur before it passes through the hot-draw bench.

15

The hot-
draw bench, as
seen in Fig. 10a
consists of a
heavy cast-steel
frame which is
provided with a
powerful hydrau-
lic cylinder and
a plunger which
operates through

Fig. 10b -After first Hot-Drawing Pass.

the full length of the bench.   The size of the
plunger may be changed to suit the desired diameter
of the tube and various sizes of dies may be drop-
ed in the recesses of the bench-frame so that the
tube will be reduced in diameter and thickness of
wall as it passes through.   In order to obtain
the desired length diameter or wall thickness, it
may be necessary to repeat the hot drawing operation several times. The head, or closed end is
then cut off and the tube is completed.

The fourth process, which is by far the most
important method of manufacturing seamless tubes
is a rotating piercing process.   It is commonly
called the "Mannesmann Process", as a German
engineer of that name made the discovery which
led to its development.

The Mannesmann process is based on the phenom-
enon of center rupture.   If a heated cylindrical
steel billet is placed on an anvil and struck a blow
with a hammer, the billet becomes elliptical in
cross section at the point of impact, the horizontal diameter having been stretched.,   If the billet is rapidly revolved about its axis as the
blows of the hammer are repeatedly delivered, each
diameter that assumes a horizontal position as the
blow is struck will be stretched and a strain will
be stretched and a strain will be set up in the
metal in the direction of every possible diameter, and as a consequence, the center of the
billet will be strained beyond the elastic limit,

16

Fig.10c. Group of Cups.

and finally a small axial hole will be torn in the billet.

The Mannesmann process performes this operation by means of rolls. The solid cylindrical billet which is heated to about 2200 deg. F. is passed between two rolls whose axes are oblique to the axis of revolution and which revolve both in the same direction at a speed of 300 R.P.M. or more. The metal of the surface of the bar thus acquires an increased motion in a spiral direction and is drawn over its core, thus making an opening and receiving the form of a pipe. The relative position of the rolls to each other and to the driving shaft is made adjustable through a large angle, and has to be regulated with the nicest precision. It is not practical without an excessive expenditure of power to make the interior diameter of such a pipe very large, but it is sufficient that an interior space is created, for there is no difficulty in widening it over a mandril.

This is done between rolls in the form of conical discs revolving in opposite directions. Since in this operation also the pipe moves spirally forward, and all its parts are spirally pushed and pressed, the metal becomes still denser. It is this spiral arrangement of the material which gives the Mannesmann pipes such remarkable strength, quite apart from the advantage they possess in presenting no line of welding. Moreover, any blow holes which are liable to be present in the metal are squeezed out spirally so as to make the walls of the tube completely impermeable. A proof of this is the retention of hydrogen for weeks in a piece of Mannesmann pipe sealed at both ends. Pipes made in this manner and enlarged, have been successfully produced in all diameters up to 16 inches.

As the rope like twist which is imparted to the metal is a very great strain on the metal, so that none but the very best grade will stand up, so several manufacturers have since sought to reduce the twist and consequently the great strain in the piercing operation.

The Steifel Piercer, patented in 1895, while similar in action and fundimentally the same as the Mannesmann machine, gives the same speed of rotation to all sections of the billet as it passes through, and consequently pierces the billet with the fibres left practically straight throughout.   In this machine discs slightly inclined to each other are used instead of conical rolls.   The practical advantages of the Stiefel Mill over the Mannesmann Mills are that the Stiefel Mill produces much less waste, is simpler in construction and requires less skill in operating.

Eight foot Riveted Steel Pipe through Ashokam Dam
Board of Water Supply, City of New York, June, 1907

Fig. 11.

Fig. 12.   66" I.D. Exhaust Steam Header for the
Milwaukee Electric Ry. Co.

- Fabricated Tubes -

In addition to the methods of manufacturing pipe by the welded or seamless process, there are a variety of other methods which may be classified under the head of fabricated tubes. These may be divided into two groups ; Those made from steel plate by forming the tube and riveting the seam, and those made from steel plate by forming the plate and interlocking the seam. In the first group is found straight riveted pipe and also a riveted pipe with a helical seam commonly known as spiral riveted pipe.

Before the manufacture of lap welded pipe had been perfected, the only method of manufacturing pipe of large diameters and for appreciable pressure was by the straight riveted process. This consists of forming the skelp into tubes and riveting the longitudinal seam usually with a double row of rivets. The pipe is made into convenient lengths for handling, usually twenty feet or thirty feet long, by riveting together either conical sections or alternately large and small cylindrical sections, each section from three to six feet long. The double rows of holes for the longitudinal seams and the single rows for the round about seams are punched by power machines and all over-lapping edges are bevel-sheared.

In forming the tube the width of sheet necessary to roll to the desired circle is found by adding to the circumference conforming to the desired inside diameter 3 1/2 times the thickness of the plate plus the lap required for riveting. The riveting is usually done in a horizontal-hydraulic or pneumatic riveter and in this way the rivet is thoroughly forced down into the holes and tightly compresses the plates against the beveled edges. For shells over 5/8 inches in thickness butt joints with either one or two cover plates are sometimes used. In this case the outside cover plates are bevel-sheared and calked on both edges.

The efficiency of riveted joints varies considerably from 40% to 65% for single riveting and from 55% to 75% for double riveting. The strength of the riveted pipe depends upon the shearing resistance of the rivets and plates and therefore to

a very great extent upon the thoroughness with which the work is performed.

Spiral Riveted Pipe is made in diameter from 3 inches to 42 inches and can be furnished in any desired length, however, the standard lengths of 20 feet or 30 feet are most generally used. In manufacturing the pipe skelp is used which is obtained from the mill in 30 foot lengths and all gauges from #20 (.0375") to #3 (1/4") for various pressures. It varies in width from 8 inches to 15 1/4 inches for different diameters of pipe.

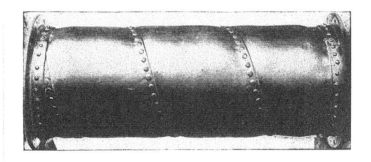

Fig. 13.

The skelp is first heated at the ends to a welding heat and welded by a special machine into long continuous strips, which are coiled on reels. It is then fed from these reels through formers into the spiral pipe machine. As the former turns and advances the pipe, it is automatically punched and riveted, the rivets being fed by hand. In passing through the former, the skelp is drawn down and lapped tightly against the preceeding circumference, the riveting then occurs, thus insuring a pressure tight seam. When the necessary length has been obtained, the pipe is cut off by means of a traveling friction saw, which is a plain disc of steel about 1/4 inch thick traveling at a high rate of speed. After having been cut in lengths, the necessary connections are placed on the ends of the pipe for joining them, or if the quite popular steel bolted connection is used, the ends are left perfectly plain.

Fig. 14.-28" Intake Main Supplying Hydro-Electric Plant
of the Homestake Mine, Lead, S. D.

The efficiency of the diagonal riveted seam
is totally different from that of the straight
seam as the helical construction gives it a decided advantage. Fig 13 shows a section of 12
inch spiral riveted pipe which has been tested
with a hydraulic pressure of 650 lbs. per sq. in.
The pipe which was manufactured by the American
Spiral Pipe Works of Chicago is made of 16 gauge
steel of 60000 lbs. Tensile strength.   The seams
are single riveted lap joints with rivets 15/64
inch in diameter when driven, and spaced about 1
inch apart.   They were driven cold.   The angle
which the seam makes with the longitudinal axis
of the pipe is 74 deg.   Before the test, the inside diameter of the pipe was 12.1 inches and the
thickness of the plate .063 inch.   After testing,
the pipe between the seams had bulged out more
than 3/4inch on the diameter, and while at the
seams the diameter remained practically constant.
The theoretical bursting strength of 12 inch pipe
of that thickness and tensile strength is 625 lbs.
per sq. in.   It is therefore seen that the test
pressure was carried beyond the point at which
the pipe might be expected to burst if it were
made of one solid piece of metal with no seams to
weaken it.

Figuring the strength theoretically, the con-
ditions of the problem are as follows: Thickness
of plate .063 inch; tensile strength of plate
60000 lbs.; diameter of rivets .2343 inch; area of
one rivet .0431 sq.in.; pitch of rivets 1 inch;
shearing strength of rivets 45000 lbs. per sq. in.;
angle of inclination of seam with girth seam 16 de g.
or 74 deg. with axis of pipe.

The efficiency of the net section of plate in
the seam is $\frac{1-.2343}{1} =$ 76.57 per cent.   The ratio
of the strength of a diagonal joint to that of a
longitudinal joint equals $\dfrac{2}{\sqrt{3 \times (\text{cos. of angle of seam})^2+1}}$

$$\sqrt{\frac{2}{3 \times \cos.^2 74a + 1}} = \frac{2}{3 \times (.2756)^2 + 1} =$$

1.76.   Therefore the diagonal seam is 1.76 times

24

Fig. 15. - Pipe line crossing R.R. track on trestle.

as strong as the longitudinal seam and its efficiency is 1.76 x 76.57 = 135; that is the
Joint is 1.35 times as strong as the solid shell
of pipe.

Figuring now on the strength of rivets in the
seam by referring to fig.15a it is seen that the
pitch in a longitudinal direction is equal to the
pitch of the rivets along the seam times the sine
of the angle, 16 deg. This gives the longitudinal

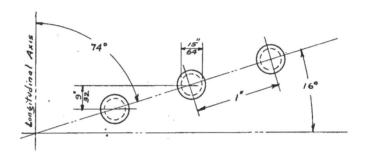

Fig.15a   Dimensions of Diagonal Seam.

pitch to be .2756 inch or about 9/32 inch.   The
strength of a solid section of plate for this
pitch is 60000 x .2756 x .063 = 1042 lbs.   The
strength of the rivets for one pitch is .0431 x
45000 = 1940 lbs.   Therefore, the efficiency of
the rivet is $\frac{1940 \times 100}{1042}$ = 186 percent, and the
strength of the rivets in the seam is 1.86 times
as great as the solid plate.

From the above calculation it is seen that a
cylindrical pipe formed with a spiral seam is
stiffer and stronger than a pipe having no Joint
whatever, as the lap reinforces the pipe and the
riveting does not lower its efficiency.

26

In the second group of fabricated tubes or
those made from steel plate rolled into a cylinder
with an interlocking seam, may be found the so called
lock bar pipe, which is used for high as well as
low pressures, and straight lock seam and spiral
lock seam pipes which are not recommended for
pressures over 125 lbs. per sq. in.

Look bar steel pipe is made by rolling two
sheets of steel into half cylinders and joining
them by means of a steel bar grooved on both
sides as shown in fig. 16 .   The metal at the

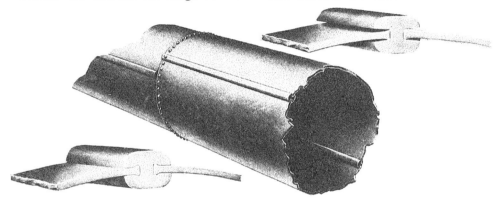

Fig 16

longitudinal edges of the sheets is upset and
thickened so that when the grooves of the lock bar
are closed upon it, the sheet is dove-tailed into
the bar.   The sheets are rolled slightly conical
and rivetedtaper joints are made at the ends to
make the pipe of a length convenient for handling.
This type of construction is very rigid, the lock
bar acting as a reinforcement to the pipe itself
to such an extent that the manufacturers claim
100% efficiency at the joint.

This pipe has found favor among many engineers,
and is used considerably for water supply lines of
some length on account of its low frictional resistance.   The table below compares it with cast
iron and riveted steel pipe.

Fig. 17. - Laying Pipes of Denver Union Water Co.

Table 1

TABLE OF DISCHARGE WITH DIFFERENT SURFACES.

Bore of Pipe assumed - - - 12 inches.
Length of Pipe assumes - 1000 feet.

Head	Kind of Pipe	Discharge Gals./sec.	Percent. Discharge.
10 ft.	Wrought-iron Welded	33.0	100
"	Riveted - - - -	27.2	82.4
"	New Cast Iron - -	27.8	84.1
"	Incrusted Cast Iron	18.6	56.4
100 ft.	Wrought-iron Welded	123.0	100.
"	Riveted - - - -	93.2	75.8
"	New Cast Iron - -	91.2	74.1
"	Incrusted Cast Iron	58.9	47.9

The most noteworthy installation of lock bar pipe is the Coolgardie water supply line in Western Australia, which is the longest steel pipe line in the world, being 351 1/2 miles long. As the question of friction was one of very great importance, a commission of eminent engineers gave very thorough consideration to the relative merits of all kinds of pipes, as a result 30 inch lock bar steel pipe was recommended and used. On the completion of the line two tests, each of twelve hours duration were made, one over 22 miles and the other over 12 miles of pipe. The result showed a frictional resistance of 2.25 feet per mile for the same discharge.

Lock seam pipe is made in several different ways. By one method the pipe is made from a single sheet, the edges being rolled over and interlocked. Another method employs the use of an additional narrow steel strip, which is placed on the outside and is interlocked withthe edges of the plate much in the same manner as the first mentioned. In each case the joint is rolled down under high pressure so that it is closed and made pressure tight. This pipe is made in gauges from #24 to #12 U.S.Standard, and will safely withstand pressures from 30 lbs. to 60 lbs. From pressures above 60 lbs and up to 150 lbs. per sq. in., depending on the diameter, a combination lock seam

29

Flow of Water in Clean Iron Pipe.   Loss of Head due to Friction

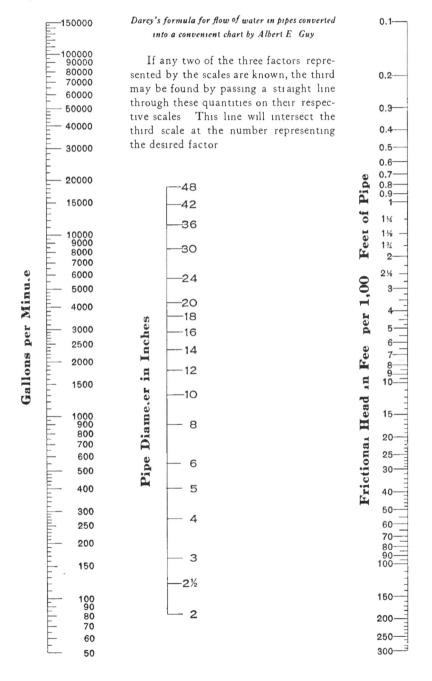

Darcy's formula for flow of water in pipes converted into a convenient chart by Albert E Guy

If any two of the three factors represented by the scales are known, the third may be found by passing a straight line through these quantities on their respective scales This line will intersect the third scale at the number representing the desired factor

and riveted pipe is used.   In this case the rivets are placed through the seam and hold it in place.

Spiral Lock Seam Pipe has the same advantage over the straight lock seam pipe that the spiral riveted has over the straight riveted pipe.   The seam, which is formed in a very similar manner to that of the straight seam, has the advantage of the helical construction, which tends to reinforce the tube and strengthen it.   As in the manufacture of spiral riveted pipe, the sheets are welded into a continuous strip and fed onto spools.   These strips, together with the narrow locking strips, which come from the mill in long lengths wound on spools, are fed together into the machine which joins and makes the finished pipe.

A piece of 18 inch #20 pipe 6 ft. 6 in. long was made the subject of a test at Armour Institute of Technology, and withstood a hydraulic pressure of 235 lbs. per sq. in., before it failed by bulging between the seams.   The seams remained tight and were not opened up or fractured in any way, thus indicating the efficiency of this type of joint.

The advantage of the lock seam over riveted pipe is the fact of the smooth interior which it has, thus reducing the frictional resistance.   In paper mill work, where the pulp flows through pipe lines, this interior smoothness is a decided advantage.

Fig.17a   Penstock , 24-inch Spiral Riveted Pipe

Peninsular Hydraulic Company.

Fig. 18   Filling In Low Ground   Sao Paulo, Brazil.

Fig. 19 - Intermediate Centrifugal Pump for Boosting
Pressure on Line mentioned above.

The Uses of Steel Pipes and Methods of
Making Connections.

Steel Pipes are used chiefly for the conveyance of materials, either, liquid, gaseous or solids
from one particular place to another.   The result
to be obtained may be merely to remove the material
as in the case of dredging lines, to furnish a
method of supply, such as a gas, water supply or
oil line, or to present a liquid or gas under pressure for the performance of work such as a compressed air line, steam or hydraulic power line.

The most suitable type of construction of the
pipe and also the method of connecting the ends depends upon the usage to which it is to be put.   In
dredging lines while the pressure is very slight,
it is necessary to have a heavy thickness of wall
as the wear on the pipe due to the friction of
the material passing through is very great.   The
connections in this case should be very strong and
capable of withstanding the excessive strains due
to the weight of the pipe and the unevenness of
the river bottom.   On the other hand an air blower
line which rarely if ever exceeds a pressure of 5
lbs. per sq. in., would be build of light gauge
material riveted and connected by slip Joints or
light flange connections.

Fig. 20

33

Table 2. - Uses of Steel Pipes .

Kind of Line	Pres. #/Sq.In.	Kind of Pipe and Connection.	
ive Steam line High Pres.	250	Wrought Steel	Flange
" " " Med. Pres.	125		
" " " Low Pres.	25		Screw Couplin
xhaust Steam Line	3	Spiral Riveted Straight "	Light F.
acuum Line	14 abs.	"	"
as Line High Pres.	100	Wrought Steel	Screwe Line Pi Coopli
" " Med. Pres.	6	Spiral Riveted Wrought Steel	Bolted Acetylene
il Line High Pres	600	Wrought Steel	Screwe ine P Coupl.
ompressed Air Line High Pres.	300	"	Flange s
" " " Med. Pres.	100	"	Screw Couplin
ir Blower Lines	5	Spiral Riveted Straight "	slip Jo.
yd. High Pres.(Hydro-Elec. Power)	500	Lap-welded Straight Riveted	Flange
" " " (Line for Hyd. Mchy)	5000	Wrought Steel	Flang
" " (Hyd. Mining)	200	Spiral Riveted Straight "	Flange Bolted.
" Med. Pres.(Town Water Supply)	75	Spiral Riveted Lock Bar	Bolted Flange
" Low Pres. (Discharge Line)	15	Spiral Riveted Straight "	Bolted or Light
" " " (Dredging Ling)	15	Welded Straight+Riveted	Flang
rine Circulating Line	15	Spiral Riveted Wrought Steel	"
aper and Pulp Line	15	Spiral Lockseam Spiral Riveted	

Note - In some cases Wood Pipe or Cast Iron Pipe are used to advantage.

Fig. 21   Two 40 inch Exhaust Lines. Chicago Railways Co.

Fig. 22 - Temporary Gas Mains - New York City.

Lead Joints.

Table 2, shows a representative list of uses
to which steel pipe is put. The most suitable
pipe for these installations of course depends on
a great many things. The location of the line
and its accessibility in regard to shipment from
the manufacturer, the ease with which the pipe
may be installed, its frictional resistance and its
durability. As each pipe line is somewhat of a
problem of its own, it is rather difficult to
specify any one particular pipe or form of connection as being the best for any pipe line in a certain class. On this account an alternative pipe
is given on the table.

Among the numerous connections for Joining
steel pipes is the Spigot and Socket Joint, which
is an adoptation of the well known Spigot and
Socket Joint used so long for cast iron pipes.

This Joint which is usually made tight by the
use of molten lead is made up in the following way.
The spigot end of one pipe is inserted into the
socket end of the next; yarn is then put into the
socket and caulked up tight. After this a collar
of clay is usually placed around the pipe closing
the end of the socket, with the exception of a
funnel shaped opening at the top into which the
molten lead is poured until the whole space between
the yarn and clay is filled. The clay is then removed and the lead caulked up hard into the socket
in the same manner as in making up lead joints with
cast iron pipes. In fixing the clay it is necessary to allow a small space around the pipe outside of the socket to allow for the compression of
the lead while caulking.

In many installations of recent years, lead
wire of a square or round cross-section has been
used instead of molten lead, and it has given ex-
celent satisfaction. It is highly recommended
by Stewarts and Lloyds of England. It is more
easily handled than molten lead and when placed
in the socket after the yarn and thoroughly caulked, the whole mass of lead is made homogeneous.
The lead wire has no tendancy to contract as the
molten lead does upon cooling, and it has been

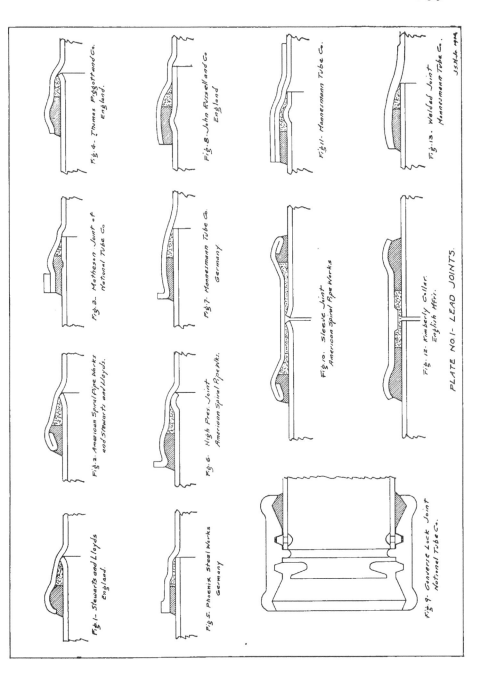

36

Fig. 1- Stewarts and Lloyds England.

Fig. 2- American Spiral Pipe Works and Stewarts and Lloyds.

Fig. 3- Matheson Joint of National Tube Co.

Fig. 4- Thomas Piggott and Co. England.

Fig. 5- Phoenix Steel Works Germany

Fig. 6- High Pres. Joint American Spiral Pipe Wks.

Fig. 7- Mannesman Tube Co. Germany

Fig. 8- John Russell and Co England

Fig. 9- Converse Lock Joint National Tube Co.

Fig. 10. Sleeve Joint American Spiral Pipe Works

Fig. 11- Mannesmann Tube Co.

Fig. 12- Kimberly Collar English Mfrs.

Fig. 13- Welded Joint Mannesmann Tube Co.

J.S.H.Jr. 1914

PLATE NO I- LEAD JOINTS.

found by cutting open sections of Joints, that those
made with the lead wire were perfectly tight
throughout while those made with molten lead and
caulked, were tight only at the face of the socket
where the caulking had penetrated. It is also
true that the Joint made with lead wire requires
only about one half the amount of lead that is
necessary when molten lead is used.

Plate 1, shows a number of typical lead joints
which are in use at the present time. They are
made by heating the end of the pipe to a bright
red heat then placing it quickly in an open end
roll and rolling it to the desired shape, or by
rolling a steel collar, and making the Joint by
placing the plain end pipe into it as is the case
with the Sleeve Joint, Fig. 10 , of the American
Spiral Pipe Works and the Kimberly Collar, Fig. 12,
manufactured by Thomas Piggott Sons and Stewart &
Lloyds of England.   The Converse Lock Joint of
the National Tube Co., Fig. 9 , is a cast iron
sleeve recessed to hold the lead, much the same way
as the others are, and in addition has two tee
shaped pockets cast in place diametrically opposite.   Two rivets placed close to the end of
the pipe engage the slopes of these wedge shaped
pockets when the pipe is inserted into the hub
and slightly rotated.   These force the ends of
the pipe against the central ring of the hub and
lock them in position ready for the lead which
makes the Joint.   For very high pressures the
Joint is reinforced with a clamp and rubber packing which increases its efficiency considerable.

The Piggott Joint, Fig. 4 , and Joints Figs. 2,
and 1 , manufactured by the American Spiral Pipe
Works of this country and Stewart & Lloyds of
England are considered the best types. A close
inspection of these Joints will reveal the following:
1st. _ That the socket end being curved inward
gives extra strength to the joint where it would
otherwise be weak and makes it impossible for the
lead to be blown out by ordinary pressure.
2nd. _ The taper sleeve, which is intended to be
filled with yarn, gives ample security that the
spigot end will not be drawn out of the socket, in
the event of subsidence of the soil in which the

38

Fig.23, 16-Inch Matheson Joint Pipe
In Colorado.

pipes may be laid.

3rd. - The spigot end being turned up engages the yarn and when a strain (due to pressure, expansion or other cause) is exerted, the yarn acts as a cushion, keeping the Joint at all times tight and secure.

4th. - The socket is made just sufficiently larger than the pipe to enable the thinest practicable caulking tool to be used, thus ensuring that only the minimum thickness and therefore the minimum amount of lead is used, as it has been found from experience that a thick band is not only a waste of lead, but a possitive weakness in the Joint.

The Matheson Joint, Fig. 3 , of the National Tube Co., is also highly recommended; a considerable amount of pipe made up with this Joint is in use in this country. It is made with a reinforcing band shrunk on the exterior of the socket which gives it the stiffness necessary at that point, as the caulking of the lead has a tendancy to open up the socket. The spigot end, however, is not quite so well taken care of as the locking is accomplished by a slight goove machined in the tube which would naturally have a tendency to weaken it.

In Joint Fig. 2 , the reinforcement of the socket is well taken care of. This is done by means of rolling the metal back on itself and giving it a double thickness at the end. This Joint, however, is not made in thickness of steel over 5/16 inch, and consequently is used only for medium and low pressures. Fig. 6 , shows a high pressure joint manufactured by the same company which is made of a single thickness of steel. The flanged-over end of the socket gives it the necessary strength, and the ball rolled on the spiggot end locks the lead, expansion being taken care of by the hemp which lies next to the lead. The form of Joint is such, that the minimum amount of lead is necessary to make it up properly.

Fig.24. 16-Inch Matheson Joint Pipe
With National Coating.

The other joints shown are similar to these types. Figs. 5,7,11,being made by the Mannesmann Tube Co., and the Phoenix Steel Wks., of Germany. The steel reinforcement bands recommended by the former and used on its joint,Fig11,is not thought to be necessary by the English and American Manufacturers. The Mannesmann Tube Co., also manufacture the joint shown in Fig. 13, which is a welded-on socket of heavier material than the pipe itself. The long sleeve joint Fig.   , is recommended where serious subsidence of the soil is anticipated or when road traffic is very heavy. These joints have the advantage that, under exceptional strains, the pipes spring, or give, without disturbing the lead.

The curves given in plate No. 5, show the comparison in weight per mile of cast iron pipe and steel pipe with the Matheson Joint. This shows the great difference in weight for the same carrying capacity. All this excess weight in cast iron pipe is due to the fact that the thickness must be greater to resist an equal internal pressure.

The welding of steel pipes into continuous mains is a method which has but recently come into use and is meeting with considerable success and favor. At the present time it is used chiefly for gas lines and supersedes the old time lead joint which has been known to open up slightly under severe strains or shocks, and cause considerable leakage and waste. Stewarts & Lloyds of England have perfected this method and do the work by means of a portable acetylene welding outfit. The acetylene generator, gas holder and purifier and the oxygen cylinders are mounted on a small carriage which can be wheeled from joint to joint along the trench, into which the pipe is laid, as the welding progresses.

The pipe ends which are welded, consist of a spigot with a slight taper which is drawn tightly into a socket with a corresponding taper as shown in Fig. 24a. The welding is then effected by fusing soft Swedish charcoal iron wire with the metal of the spigot and socket by the oxy-acetylene process. The close fit of the tapered surfaces of

spigot and socket relieves the actual weld of any
bending stresses which may come on the pipes.   In
many cases it is convenient to weld the pipe above
ground over the trench and reel it into the trench
as the welding progresses.   After welding the
joints are tested with water or compressed air by
means of a hand operated pump or a power driven
compressor depending upon the size of the pipe to
be tested.

Steel pipes of this type have been supplied
not only for gas lines, but also for water and oil
pipe lines.   In one instance an oil line was laid
under water and as the pipes were welded they were
paid out like a cable from a barge which moved
along as the welding progressed.   In another case
they were welded on the bank and drawn in one length
across a river a mile wide.   They have also been
used for pump delivery mains in vertical pit mine
shafts, the welding being done in the shaft and
sometimes on the pit-head.   In the latter cases
the main is let down a pipe at a time as each
joint is completed.   This places a very great
strain on the joints as a vertical line of pipe
1000 ft. long creates a tensile stress on the joint
of over 3000 lbs. per inch of pipe section.   This
method, however, may be used with safety for very
deep shafts, for joints have been found to stand
under test more than 40000 lbs. per square inch.

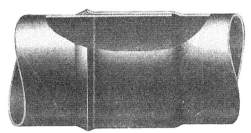

LONG SLEEVE PATENT WELDED
JOINT

Fig.24a.

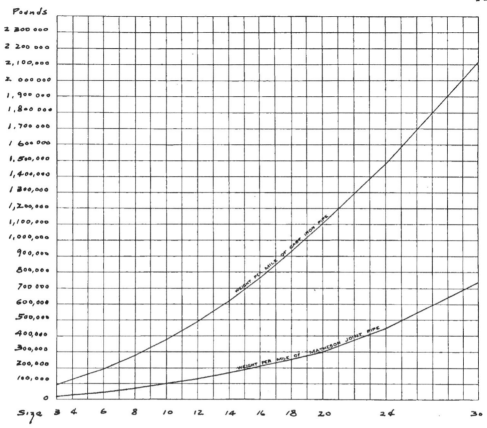

Fig.25. Curve Showing the Comparison
in Weight per Mile of Cast Iron
and Matheson Joint Pipe.

Fig. 26 – Shipment of Dredging Pipe.

Screwed Joints.

Screwed Joints are used to a great extent in interior piping up to 12 inches in diameter, with moderate pressures. Larger than this and for higher pressures, flanges or riveted joints of various types are used. Gas, compressed air, steam, oil and many other fluids are piped through wrot steel pipes with screwed connections.

Plate No. 2, shows a number of screwed connections which are used quite regularly.

The standard coupling joint Fig. 9 , is the connection generally used. Couplings for standard pipes have straight threads while the pipe threads have a taper of 3/8 inch to 1 foot, and are cut according to the well known Briggs Standard.

Robert Briggs about 1862 while Superintendant of the Pascal Iron Wks., formulated the nominal dimensions of pipe up to and including 10 inches. They are as follows: The nominal and outside diameters and pitch of thread are given in the following table. The dimensions up to 10 inches are Briggs figures and the others were added to his table.

The thread has an angle of 60 deg., and is slightly rounded off at top and bottom, so that the total height (or depth), $H = \frac{0.8}{n}$, – n being the number of threads per inch. The pitch of the threads $\frac{1}{n}$, increases roughly with the diameter, but in an irregular manner, It would be of advantage to change the pitches except for the fact that they are now firmly established. Fig.27 shows a section of pipe threaded according to Briggs Standard. The conical threaded ends of pipe are cut at a taper of 3/8 inch per foot of length (I.E. 1 in. 32 to the axis of the pipe)

46

The thread is perfect for a distance   L from
the end of the pipe, expressed by the rule
L $\dfrac{0.8\ D + 4.8}{n}$;   where D = outside diameter in
inches.   Then come two threads, perfect at the

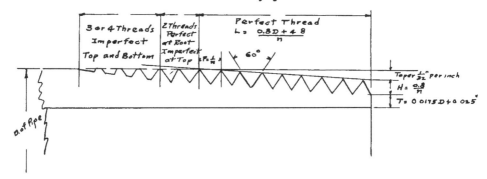

Fig.27. Section of Briggs Standard Thread.

bottom, but imperfect at the top and finally come
three or four threads imperfect at both top and
bottom.   These last do not enter into the joint
at all, but are incident to the process of cutting
the threads.   The thickness of the pipe under the
root of the thread at the end of the pipe equals
T = 0.0175 D + 0.025 inches.

After screwing together a number of standard
pipes, it will be found that at nearly every joint
a portion of each pipe thread remains exposed outside of the socket.   These are the weak portions
of the pipe, and there is always a danger of breakage at the bottom of an exposed thread from bending
stresses which cannot always be avoided in laying
a line of pipe.

The line pipe coupling Fig. 10 , is a modified
form of the standard coupling, from which it dif-
fers in the following important details.
    1st. It is longer and heavier.
    2nd. The ends are recessed, in order that they
may fit the pipe snugly just outside the thread,
which is thereby fully protected from any bending
stresses that may come apon the pipe.
    3rd. The threads have a taper of 3/8 inch to 1
foot, to correspond to the taper of the thread of
the pipe.   This insures a perfect contact for every

l

47

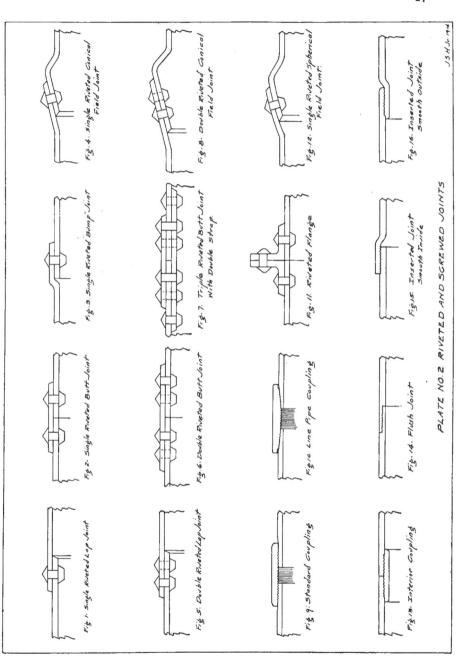

Fig.1. Single Riveted Lap Joint

Fig.2. Single Riveted Butt Joint

Fig.3. Single Riveted Bump Joint

Fig.4. Single Riveted Conical Field Joint

Fig.5. Double Riveted Lap Joint

Fig.6. Double Riveted Butt Joint

Fig.7. Triple Riveted Butt Joint With Double Strap

Fig.8. Double Riveted Conical Field Joint

Fig.9. Standard Coupling

Fig.10. Line Pipe Coupling

Fig.11. Riveted Flange

Fig.12. Single Riveted Spherical Field Joint

Fig.13. Interior Coupling

Fig.14. Flush Joint

Fig.15. Inserted Joint Smooth Inside

Fig.16. Inserted Joint Smooth Outside

PLATE NO.2 RIVETED AND SCREWED JOINTS

JSH Jr.1914

	Diameters			Wt.per foot		Threads per inch		Couplings		
	External	Internal	Thickness	Plain ends	Threads and Couplings	Threads per inch		Diameter	Length	Width
8	.405	.269	.068	.244	.245	27		.562	7/8	.029
4	.540	.364	.088	.424	.425	18		.685	1	.043
8	.675	.493	.091	.567	.568	18		.848	1 1/8	.070
2	.840	.622	.109	.850	.852	14		1.024	1 3/8	.116
4	1.050	.824	.113	1.130	1.134	14		1.281	1.5/8	.209
	1.315	1.049	.133	1.678	1.684	11	1/2	1.576	1 7/8	.343
4	1.660	1.380	.140	2.272	2.281	11	1/2	1.950	2 1/8	.535
2	1.900	1.610	.145	2.717	2.731	11	1/2	2.218	2 3/8	.743
	2.375	2.067	.154	3.652	3.678	11	1/2	2.760	2 5/8	1.208
2	2.875	2.469	.203	5.793	5.819	8		3.276	2 7/8	1.720
	3.500	3.068	.216	7.575	7.616	8		3.948	3 1/8	2.498
2	4.000	3.548	.226	9.109	9.202	8		4.591	3 5/8	4.241
	4.500	4.026	.237	10.790	10.889	8		5.091	3 5/8	4.741
2	5.00	4.506	.247	12.538	12.642	8		5.591	3 5/8	5.241
	5.563	5.047	.258	14.617	14.810	8		6.296	4 1/8	8.091
	6.625	6.065	.280	18.974	19.185	8		7.358	4 1/8	9.554
	7.625	7.023	.301	23.544	23.769	8		8.358	4 1/8	10.932
	8.625	8.071	.277	24.696	25.000	8		9.358	4 5/8	13.905
	8.625	7.981	.322	28.554	28.809	8		9.358	4 5/8	13.905
	9.625	8.941	.342	33.907	34.188	8		10.358	5 1/8	17.236
	10.750	10.192	.279	31.201	32.000	8		11.721	6 1/8	29.877
	10.750	10.136	.307	34.240	35.000	8		11.721	6 1/8	29.877
	10.750	10.020	.365	40.483	41.132			11.721	6 1/8	29.877
	11.750	11.000	.375	45.557	46.247			12.721	6 1/8	32.550
	12.750	12.090	.330	43.773	45.000			13.958	6 1/8	43.098
	12.750	12.000	.375	49.562	50.706			13.958	6 1/8	43.098
	14.000	13.250	.375	54.568	55.824			15.208	6 1/8	47.152
	15.000	14.250	.375	58.573	60.375	8		16.446	6 1/8	59.493
	16.000	15.250	.375	62.579	64.500	8		17.446	6 1/8	63.294

49

Fig. 28 - Building the Dam of the Johnstown Water Co

thread, a prime essential for tight joints.

A leaky line pipe joint indicates imperfect or damaged threads, or carelessness in connecting the pipes. To avoid damage from transportation, it is the practice of the best mills to screw a heavy guard or protector on the exposed thread of each length of pipe.

In California, line pipe is largely used to convey oil under pressure. A considerable number of 2 inch, 3 inch, and 4 inch lines are in operation in the several oil fields of the State, and some of them are subjected to very high working pressures. The Standard Oil Companies 8 inch line extending from the Bakersfield and Coalinga oil fields to Point Richmond on San Francisco Bay, a distance of 278 miles was completed in 1903. It is used as a pumping main for the transportation of oil and carries a working pressure of approximately 600 lbs. per sq. in. Before laying, each pipe was tested at the mill to a hydraulic pressure of 1500 lbs. per sq. in.

Line pipe is also used in California for the transmission of gas under high pressures. In many localities it is more economical to supply two or more towns from one source, through small pipes, at high pressures, than to construct and operate a generating plant in each town. In one of the first attempts at high pressure gas transmission in California, 2 inch standard pipe was used, but after the completion of the line, the joints leaked so badly at a pressure of 15 lbs. per sq. in., that it became necessary to take up the pipe and relay the entire line, replacing the couplings with line pipe couplings. Another line, 9 miles in length of 2 inch and 2 1/2 inch line pipe was laid with great care, precaution being taken to test the line at frequent intervals during the progress of the work. Upon the completion of the first 5 miles of the line, it was tested to 100 lbs. air pressure for 36 hours, without developing the slightest leak. These examples show the superiority of line pipe couplings and the advis-

ability of using them in preference to standard couplings for high pressures.

The Flush Joint and Inserted Joints shown in Figs. 14, 15 &16, used principally in connections with boring tubes and well casing.

Fig. 29   Brine Circulation Line

52

Riveted Joints.

Riveted joints are frequently used in pipe lines whose inside diameters are not less than 20 inches, this being the smallest pipe in which even an under sized riveting helper can work to advantage. The maximum head for which this type of joint should be used is about 1200 ft. (530 lbs. per. sq. in.), although it has been used for even higher heads. Figures 1 to 8 Plate No. 2 show the various types of lap and butt joints which are used, depending upon the pressure on the line.

The making up of riveted connections in the field has to be done with considerable care and by experienced workmen. The trouble experienced with the pressure line of the Central Colorado Power Co., shows this very plainly. The Boulder pressure line of the Central Colorado Power Co., is a steel plate straight riveted pipe about 9500 ft. long, and varying in inside diameter from 44 inches to 56 inches. The static head is approximately 1800 ft. The light portion of the pipe is lap riveted at both longitudinal and girth seams while the heavy portion of the pipe which has a slope length of 4450 ft. and weighs about 1500 tons is butt riveted with a longitudinal joint efficiency of about 80%. The slope on which the pipe is laid varies from 5 to 45 deg. a good part of the butt strap pipe laying on slopes of about 30 deg., The line was laid in a trench with the top of the pipe about 3 ft. below the original surface, the foundation being either solid or disintegrated rock.

Beginning with the first field test of the line, trouble was experienced from leakage in the riveted pipe. After various attempts to stop the leaks by calking and other methods, success was attained by welding with an oxy-acetylene torch. The leakage was found to be most serious at the field joints of the butt riveted pipe. These joints had an outside and an inside cover strap similar to Fig. 7 the former being double riveted and the latter triple riveted while the girth seams were single riveted. The plate in the butted pipe varied from 1/2 inch to 1 3/4 inch thick and the leakage occured principally where the longitudinal and girth seams over lapped. Thermit was tried to repair the

53

Fig. 30 -36 inch Straight Riveted Pipe of
the Rio De Janeiro Power Co.,

joints but could not be used because of the difficulty in securing the mold so that it would
contain the molten metal during the process of
combustion and at the same time not injure the
pipe.

In laying riveted pipe, the lengths are
riveted together after the manner of connecting the
short sections in the shop, each length having a
large and a small end.   Field riveting and caulking are sometimes done by hand and sometimes by
compressed air.   Riveted joints are also used for
welded pipes.   They are then termed "Bump" or expanded joints, Fig. 3 , because one end of each
pipe is upset or expanded.   The expanded end is
beveled for caulking.   Ordinarily the joints are
single riveted, but when very high heads are used
requiring heavy pipe, it has been found necessary
to double rivet the joints in order to make them
tight.

Bump Joints are used for high and low pressure
but are expensive and laborious to install.   They
will not take up expansion and contraction, so separate expansion joints are necessary.   The ends
of the pipes of large diameter cannot be made
with great accuracy as to size, and as sagging of
the pipe by its own weight and the shocks incident
to transportation tend to deform the ends of the
pipe from the circle, some clearance must be al-
lowed in making the bell ends.   This clearance is
detrimental to a good job of riveting, and caulking, and often required special appliances in installation.

The objectionable features are overcome in
the conical field riveted joint shown in Fig.4,
both of its ends are shaped to a conical flare so
that any deformation of the ends is held to have
no detrimental effect, on account of the self-
centering action of the cone surfaces.   The
lengths which are to go together in the field, are
made tight in the shop, and drilled for riveting.
The rivet heads cause little, if any contraction
in the pipe area.   Where necessary to deflect the
line slightly, to take care of irregulations in
the line, joint Fig. 12, is used.   This is made

55

Fig. 31 – Power House and Pipe Lines of the Rio
        De Janeiro Tramway Light & Power Co.,
        installation.

in a spherical instead of a conical shape. In
this case the holes are drilled in the field.
This joint as well as the one previously mentioned is used for very high pressure work.

Steel flanges are sometimes shrunk on the
pipe and then riveted to the pipe and to each
other. No gasket is used in this case, the
joint being made tight by caulking the beveled
edges of the flanges. Fig. 11.

Fig.31a. 48 inch Riveted Steel Conduits,
City of Newark, N.J.

Fig. 32   Producer Gas Piping in the plant of the Ford
           Motor Co., Detroit, Mich.

Fig. 33 - Producer Gas Main temporarily installed.

Bolted Joints.

Bolted Joints which is the term applied to
bolted collar connections, have been found to be
a very desirable form of connection, for many
purposes.   Plate No. 3, shows the types of
bolted joints which are manufactured by various
companies.   Figs. 1, 2, and 3, are the most
representative ones.   The Dresser Joints, two
views of which are shown in Figs. 2 and 3, are installed with great profusion all over the country.
The first is known as the all steel coupling, the
body being rolled and welded and the flanges forged, while the second, the Smith design is equipped
with a steel body and malleable iron flanges.
These are manufactured for plain or threaded pipe
from 6 inches to 20 inches or more.

The joint manufactured by the American Spiral
Pipe Works and shown in Fig. 1 is all steel also.
It has been installed in pipe lines up to 42 inches
in diameter, and in many lines under high heads
and trying circumstances it has given perfect satisfaction.   It is quite similar to the Dresser Joint,
the flanges in this case being made of steel
angles.   These flanges when drawn up tightly by
the bolts, compress the rubber packing against the
body and make the joint secure.   One great advantage of the Bolted Joint is the fact that it
forms a perfect expansion joint.   Bends may be
made with it also, as it is possible to make a
slight deflection of the line at each joint.

The Slip Joint shown in Fig. 4, is used largely
for medium and low pressure work.   The sleeve
which is attached to one end of the pipe is wrapp-
ed with burlap or canvas soaked in red lead or
liquid asphaltum, and then driven into the adjoin-
ing pipe.   The lugs on the exterior are then
wired together in order to make the pipe secure.
Another type of slip joint which is used for large
diameter pipes is shown in Fig. 6.   It has lugs
which are bolted together.

The Bolted Socket or Submarine Joint, Figs. 7
and 8, is especially suited for long line work, or
for connections on submerged pipe lines.   These

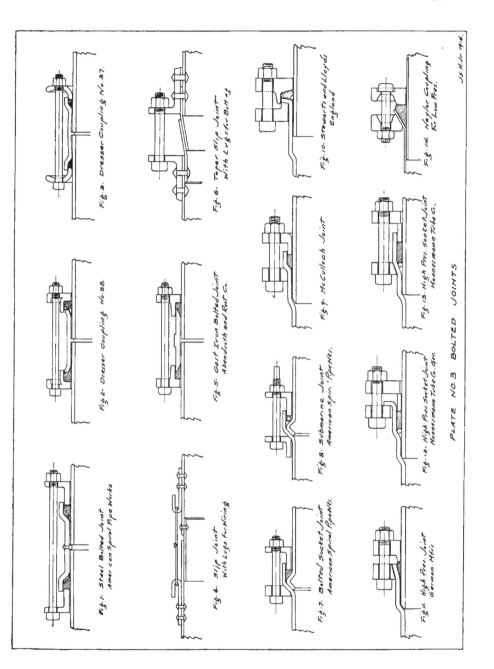

59

Fig.1. Steel Bolted Joint
Amer con Spiral Pipe Works

Fig.2. Dresser Coupling No.38.

Fig.3. Dresser Coupling No.37.

Fig.4. Slip Joint
With Lugs for lining

Fig.5. Cast Iron Bolted Joint
Abendroth and Root Co.

Fig.6. Taper Slip Joint
With Lugs for Bolts

Fig.7. Bolted Socket Joint
American Spiral Pipe Mfr.

Fig.8. Submarine Joint
American Spir. Pipe Mfr.

Fig.9. McCulloch Joint

Fig.10. Stewart and Lloyds
England

Fig.11. High Pres. Joint
German Mfr.

Fig.12. High Pres. Socket Joint
Mannesmann Tube Co. Ger.

Fig.13. High Pres. Socket Joint
Mannesmann Tube Co.

Fig.14. Naylor Coupling
Fr. Low Pres.

Js. H. L. 1916.

PLATE NO.3 BOLTED JOINTS

joints are usually made on lap welded pipe and are
made by heating and rolling the ends.   Lead or
rubber packing is used for making the joint tight.
Figs. 9, 11, 12 and 13 are other similar types and
are adapted to the highest service pressures and
to pipe thicknesses up to 1 1/4 inch plate.   The
gaskets can be removed without taking the lead
apart.   These joints also take care of expansion
and contraction and small angular deflections.   A
modification of joint Fig. 11 is made with spherical
ends to allow large angular deflection.   Then no
expansion is taken care of and it is necessary to
insert one of the other type to allow for expansion.

The Naylor Joint, Fig. 14, is used only for
low pressure work.   It is manufactured by Robert-
son Brothers, Chicago, and is used in connection
with their straight lock seam pipe.   Its construction is similar to that of Fig. 11, and in
the proper thickness of metal would be good for
high pressures.

Fig.33a. Hydraulic Giant Washing Peat for Making Paper

Fig. 34-30 ınch Lapewelded Pipe in Power
Line of Homestake Mine.

Flanged Joints.

Flanged Joints are considered the correct
type of joint for high pressure work. Plate No. 4
shows a number of the standard types which are
manufactured the world over. The welded type of
flanged connection shown in Figs. 1 and 2, is
coming more and more into favor among engineers.
It is used for a great variety of work. The
loose rings shown in the first figure are found to
be a great advantage especially in the laying of
pipe as the bolt holes are lined up more easily.
These rings and nearly all flanges now used for
high class work are made of forged or rolled steel.
A distinctive feature shown in joint Fig. 1, is
the annular groove, into which the circular rubber
gasket is compressed when the flanges are drawn
together. No matter how great the pressure, the
gasket cannot be blown out, since the tendancy is
to squeeze it more tightly into the groove.

This style of joint was adopted for the 5 inch
pipe line in the Simplon Tunnel, Switzerland, operating under a maximum pressure of 2250 lbs. per sq. in.
A similar joint is used in a power line near Voury,
Switzerland, under a head of 3117 feet (1250 lbs.
per sq. in.)

The joint manufactured by the Mannesmann Tube
Co., and shown in Fig. 6, has proven to be a perfect
joint for very high pressure work. A great number
of tests, up to 15000 lbs. per sq. in., made to
ascertain the resistance of this joint have shown
that it does not become deformed and that by con-
tinuing the pressure the pipes break rather than
the joints. As jointing material, according to
circumstances, rings of gutta percha, india-rubber
or some similar substance are used. These rings
are prevented from giving way by a copper ring of
cross-shaped section, which, united with the
groove in the ring of the double flange, ensures
perfect tightness.

In making up the joint shown in Fig. 8, grooves
about 1/32 inch deep are machined in the pipe, and
corresponding projections are machined in the
flanges. The flanges are then heated and shrunk

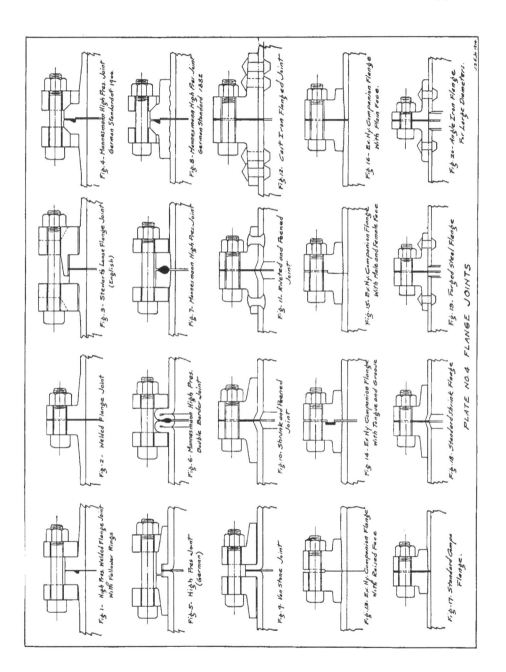

63

Fig. 4 - Mannesmann High Pres. Joint. German Standard of 1900.

Fig. 8 - Mannesmann High Pres. Joint. German Standard 1882.

Fig. 12 - Cast Iron Flanged Joint.

Fig. 16 - Ex. Hy. Companion Flange With Plain Face.

Fig. 20 - Angle Iron Flange For Large Diameters.

Fig. 3 - Stewarts Loose Flange Joint (English)

Fig. 7 - Mannesmann High Pres. Joint.

Fig. 11 - Riveted and Peened Joint

Fig. 15 - Ex. Hy. Companion Flange With Male and Female Face.

Fig. 19 - Forged Steel Flange

Fig. 2 - Welded Flange Joint

Fig. 6 - Mannesmann High Pres. Joint. Double Border Joint.

Fig. 10 - Shrunk and Peened Joint

Fig. 14 - Ex. Hy. Companion Flange With Tongue and Groove

Fig. 18 - Standard Shrink Flange

Fig. 1 - High Pres. Welded Flange Joint With Follower Rings

Fig. 5 - High Pres. Joint (German)

Fig. 9 - Van Stone Joint

Fig. 13 - Ex. Hy. Companion Flange With Raised Face.

Fig. 17 - Standard Comp. Flange.

PLATE NO.4 FLANGE JOINTS

onto the pipe with the projections fitting into the recesses, the flange having been made to the correct inside diameter so that a tight fit is insured.

The Van Stone Joint, with the extra heavy high hub flange, as shown in Fig. 9, has met with great favor for high pressure work and is used extensively in the large power houses of this country. The joint is made by heating the end of the pipe and turning it over the flange, and then facing the end of the pipe. With this joint there is no possible place for leakage, except through the gasket, and this is readily taken care of. The flange is loose on the pipe and can be turned to any desired position for the alignment of the bolt-holes.

The Shrunk and Peened Joint is also a very satisfactory one (Fig. 10.) This joint is made by boring the flange a little smaller than the outside diameter of the pipe and then shrinking it on while hot. The end of the pipe is usually peened into the flange by a hand hammer or expanding machine. Some engineers require the larger size riveted in addition to the peening (Fig. 11). There is considerable strain put upon the flange by the shrinking effect and any other than forged steel flanges are unsafe for this work. This joint is also the standard for the U.S. Navy work and is used by practically all of the shipbuilders.

The screwed flange connections are used extensively for steam pressures up to 250 lbs. per sq. in. In the general use of the screwed flange, there are many modifications; some engineers prefer a raised face, others a male and female face or a tongue and groove, as shown in Figs. 13 to 16, in some cases the back of the hub is caulked around the pipe. This no doubt is a great advantage in cases where it is necessary to tighten the flange after it is erected in place. When the flanges are attached in the shop, the customary method is to screw the pipe through the flange until it projects beyond the face and then finish pipe a d flange together. In this manner the end of the

65

Fig.35 - Hydraulic Lines entering Power Hour of the
Homestake Mine.

pipe engages the gasket and makes a perfect joint.
The flanges are threaded according to Briggs Standard gauges, as the pipe is turned out, threaded
according to that standard.

The Standard Companion Flanges, shown in
Fig. 17, are of Forged Steel, and made according
to the dimensions of standard cast iron flanges
as adopted by the American Society of Mechanical
Engineers. They are sufficiently strong to
allow their being used for high pressures and in
many cases may be used instead of the extra heavy
standard.

Cast Iron and Cast Steel flanges are still
made up in considerable quantities, but are fast
going into disfavor. They are not reliable and
frequently cause trouble by breakage, through
sudden shock or over strain. Forged Steel Flanges may be securely attached to the pipe without
the possibility of breaking. This insures an
absolutely tight joint, which is very essential in
good construction. They cannot be broken in
transportation as cast iron flanges, frequently are,
and they cannot be cracked or broken in erection,
however, roughly handled, on account of their
toughness and elasticity.

Flanges made of steel angles rolled up and
welded are used to a considerable extent, especially
for large pipe lines. They are riveted to the
pipe as shown in Fig. 20, and the pipe is then
caulked against the flange on the inside. The
flanged connections shown in Fig. 19, are made in
varying thickness according to the pressure. They
are commonly used on hydraulic lines under pressure
of 350 lbs. per sq. in.

67

Testing of Pipes.

Pipes which are tested in the mill, before
shipment are usually tested by means of hydraulic
pressure.   Fig.   , shows the pipe testing machine
of the American Spiral Pipe Works.   It consists of
two heavy heads, one stationary and the other

Fig. 36.

movable.   The latter is easily moved by hand, by
means of a crank and may be locked in any position,
up to 40 feet from the other head, by means of three
sliding bolts which fit the grooves in the large
horizontal tension bolts of the machine.

The movable head has a stationary platen
fastened to it with concentric grooves filled with
packing to take the pipe end and also an outlet opening which may be closed by a valve when all the air
has been expelled from the pipe being tested.   The
stationary head has a movable platen which also
has the packed concentric grooves and is operated
by means of hydraulic pressure.   It has a move-
ment of about 15 inches to take care of intermediate
lengths of pipe as the lock grooves on the tension
rods are spaced 1 foot apart.

The pipe being placed in position, the movable
head of the machine locked and the movable platen
forced tightly against the pipe, it is filled
through the center of the stationary head by means

68

Fig. 37- Lap-welded Tube 11 feet in diameter.

Fig. 38 - Lap-welded Pipe in Testing Machine.

of a centrifugal pump, which draws the water from
the pit below.   When the pipe is full of water
and all the air expelled, the pressure may be
brought to 650 lbs. per sq. in. by means of a connection to a high pressure hydraulic line or to
2000 lbs. per sq. in. by means of a high pressure
steam pump which is located at the machine.   The
end pressure on the pipe is increased as the internal pressure is increased, and is maintained at
about one half the internal pressure.   Pipes
are sometimes tested by bolting blind flanges on
the ends, then filling them with water and imposing
the pressure by means of a high pressure pump.
This way is necessarily very slow, however, and
has no advantage over the other method.

Each length of pipe is tested to 50% more than
the specified working pressure unless a request is
made otherwise.   While under test, the pipe is
examined carefully to see if any leaks have developed.   Boiler tubes and other pipes of small diameter are tested in a machine of much smaller size
than the one first mentioned, but the same principle
is involved in its operation.

Pipe lines are tested in the field after
their installation by means of water or compressed
air with either hand operated or power driven com-
pressors.

The failure of steel pipes may be due to one
or more of a number of causes, vis: - Excessive
bursting strain, due to a shock on the line;
failure of a joint or seam, due to poor workman-
ship or inferior grade of materials; excessive
collapsing strain, due to heavy earth and rock
fill or a giving away of the supports accross a
depression, also the sudden emptying of a hydraulic
line without proper air-inlet relief valves.
Failure may also be caused by the deteriation of
the pipe line, due to corrosion and rusting away
both on the inside and outside surfaces.
Electrolysis which is considered a type of
corrosion has caused much trouble in districts
where stray electrical currents are flowing.   The
steel pipe line naturally attracts the stray currents as it is so much better a conductor than the
earth.   At the point where the current leaves the

70

Fig. 39 — 30 inch Spiral Riveted Pipe of Beaver
River Power Co.

pipe to return to the nearest power house, as well as where it enters the pipe, it seems to cause a pitting of the steel. This in time is apt to cause a dangerous break in the pipe line.

In order to protect the pipes from corrosion and rusting, it is customary to apply some kind of a coating to completely cover their surfaces, For exposed lines the pipes are usually either galvanized or painted with some kind of a mineral water-proof paint, and in some cases they are painted after galvanizing. For underground lines an asphaltum or mineral rubber coating is used to a very great extent. This is made from Gilsonite which is mined in this country in the State of Utah, it is practically a pure hydro-Carbon and is considered the best preservative for steel. Its physical properties are such that it does not melt or run in extremely hot weather, nor does it become brittle or crack in cold weather.

On account of the rough usage which the pipes ofttimes receive in transit from the mill to the place of construction, a protection to the asphaltum coating is sometimes wound on the pipe in the shape of strips of canvas or burlap soaked in asphaltum. This protects the coating from being bruised and scraped off and makes positive the fact that the pipe is well protected.

Some manufacturers roll the pipe in sand after dipping in the asphalt kettle in order to afford protection to the coating.

Fig. 40- ⎺8" and 42" Water Supply Line of Yukor. Gold Co.

Some Pipe-line Installations.

The following gives a brief description of
a number of pipe lines which may be of more than
passing interest.

The hydro-electric power plant of the Rio
De Janeiro Tramway Light & Power Co., in Brazil,
S. A., involved two steel penstocks 8 feet in
diameter by about 6000 feet in length, these branch-
ing into five feed water pipes 3 ft. in diameter
by about 500 feet to the power house.   The water
is delivered through these at a pressure of 450
lbs. per sq. in., and developes 52200 horse power.
The pipes installed were straight riveted and of
varying thickness and type of joints along the
line as the pressure changed.   The entire installation including the power house and the
transmission line was installed by the Riter-
Conley Manufacturing Co., of Pittsburg, Penn.,
Fig. 31 , shows the power house and Fig.30, a
section of the 36 inch line.

The illustration Fig. 2 , shows a portion of
a Norweigian pipe line constructed of three lines
of Mannesman Steel Tubes totaling 5300 feet.   The
lines are 48 inch, 54 inch and 60 inch, and the
maximum working pressure is 640 lbs. per sq. in.

The Coolgardie pipe line mentioned previously,
carries the water supply from the Helena River, a
distance of 351 1/2 miles to the Coolgardie Gold
Fields of Western Australia.   It is operated as
a force main and is made up of nine sections,
eight of which have their own pumping stations
while through the last the water flows 44 miles
by gravity.   In the length of the line there are
eleven reservoirs which hold from 1 million to
12 million gallows each.   The impounding reservoir
from which the water is first drawn has a capacity
of 5500 million gallons.   Lock bar steel pipe was
used entirely on this line on account of its
lightness, strength and low frictional resistance.

Fig.41-Operations of the Totok Mining Co.,Dutch East Indies.

The San Joaquin Electric Company's pipe line
(near Fresno) has the distinction of being the
pioneer high-pressure power line of the Pacific
Coast.   It was constructed in 1896.   Its length
is 4030 feet and its total head is 1406 feet.
There are 960 feet 24 inch riveted pipe No. 12
gauge steel, 860 feet 24 inch riveted 1/4 inch
steel, 400 feet 20 inch lap-welded 5/16 inch steel
with Converse joints, 800 feet 20 inch lap-welded
5/16 inch steel with flange joints, and 1000 feet
30 inch lap-welded 3/8 inch steel with flange
joints.   The flanges were shrunk on and riveted
to the pipes, one of each pair being recessed,
while the other has a corresponding annular pro-
jection.   Each joint contains 16 bolts 1 inch in
diameter.   A rubber gasket was used between the
faces.

During the year 1900 the Standard Electric
Company constructed two parallel pipe lines for
power development, each consisting of 2813 feet
48 inch wooden stave pipe, 464 feet 48 inch
riveted pipe 5/16 inch steel, 760 feet 30 inch
cast iron pipe with shells 1 inch, 1 1/4 inches
and 1 1/2 inches thick, corresponding to 275 feet,
550 feet and 700 feet static heads, respectively,
and 2365 feet 30 inch lap-welded steel pipe with
shells 7/16 inch, 1/2 inch, 5/8 inch and 3/4 inch
thick, depending upon the static head.   The total
head is 1475 feet.   The joints for all of the
lap welded pipe are of the solid welded flange
type.   The flanges are 2 1/4 inches thick.   Each
joint contains 32 bolts, 1 inch, 1 1/8 inches and
1 1/4 inches, the size depending on the pressure.

Fig. 42-Washing down Gravel at the American Hill
Mine of Yukon Gold Co.

In the East, one of the most important installations of riveted steel water pipe is that
of the East Jersey Water Co., which supplies the
city of Newark.   The contract provided for a
maximum high service supply of 25,000,000 gallons
daily.   In this case 21 miles of 48 inch pipe
was laid, some of it under 340 feet head.   The
plates from which the pipe is made are about 13
feet long by 7 feet wide, open-hearth steel.
Four plates are used to make one section of pipe
about 27 feet long.   The pipe is riveted longitudinal-
ly with a double row, and at the end joints with
a single row of rivets of varying diameter, corresponding to the thickness of the steel plates.

Before being rolled into the trench, two of
the 27 feet lengths are riveted together, thus
diminishing still further the number of joints
to be made in the trench and the extra excavation
to give room for jointing.   The thickness of the
plates varies with the pressure, but only three
thicknesses are used, 1/4, 5/16 and 3/8 inches,
the pipe made of these thicknesses having a weight
of 160, 185 and 225 lbs. per foot, respectively.
At the works all the pipe was tested to pressure
1 1/2 times that to which it is to be   subjected
when in place.

At the Mannesmann Works at Komotau, Hungary,
more than 600 tons or 25 miles of 3 inch and 4 inch
tubes averaging 1/4 inch in thickness have been
successfully tested to a pressure of 2000 lbs. per
sq. in.   These tubes were intended for a high-
pressure water-main in a Chilian nitrate district.
This great tensile strength is probably due to the
fact that, in addition to being much more worked
than most metal, the fibres of the metal run spirally, as has been proved by microscopic examination.
While cast-iron tubes will hardly stand more than
200 lbs. per sq. in., and welded tubes are not safe
above 1000 lbs. per sq. in., the Mannesmann tube
easily withstands 2000 lbs. per sq. in.   The length
up to which they can be readily made is shown by
the fact that a coil of 3-inch tube 70 feet long was
made recently.

78

Fig.43, shows the transportation of some 30 inch lap-welded steel pipes in Ceylon. They are for a main 23 miles in length for the municipality of Colombo, Ceylon, and are furnished by Stewarts and Lloyds of England. Stewarts Lead Joint, shown on Plate 1, Fig.1, is the connection used. The pipes were tested to withstand a continuous hydraulic pressure of 350 lbs. per sq.in.

Fig. 43. Transporting 30" Lap-Welded P pe in Ceylon.

Steel pipes while primarily used for the purposes of conveyance are occasionally resorted to for other uses of an entirely different nature. Stewarts and Lloyds of England have designed and are manufacturing a steel tubular truss. It is a modification of the Fink truss, and is used in combination with pipe columns for building construction. The trusses are so designed that all members in compression or subject to bending loads in two or more directions are of tubular form, while all tension members are solid round bars. They thus provide the requisite strength and rigidity with a minimum weight. Pipes are used for purlins and rolled sections for the gutters and ridge-roll. These are all connected in place by cast fittings which are screwed onto the pipe. The trusses are designed to give a factor of safety of 3 on the dead loads, including corrugated sheeting plus a horizontal wind load of 20 lbs. per square foot.

Another example of this kind, which is quite unique is the use which was made of 3 inch Spiral Riveted Pipe for building roof trusses for a mill building at the Incaoro Mine, La Pas, Bolivia, South America. The trusses were of the Fink type and took a 42 foot span. The side members and the bottom chord were made of 3 inch #20 Spiral Pipe with steel flanged connections, while the interior struts and tension members were made of 1 inch wrought pipe and 5/8 inch diameter steel rods, respectively. A 27° cast iron flanged elbow and two cast iron flanged laterals were used at the apex and the two sides of the truss. Wood purlins were bolted to the flanges at intervals of about 5 feet to support the roof.

Dredges for use in gold mining have also been built of Spiral Riveted Pipe. One partucular instance is a dredge built by the Clark Dredge Company and operated on the Saskatchewan River, Canada. It was 35 feet wide by 100 feet long, made up of 12 inch #16 Spiral Riveted Flanged Pipe. Six inch channels locked to steel plates bolted between the flanges, tied the raft together sideways. The dipper arm was also constructed of Spiral Pipe, four pipes wide by 20 feet long. The buoyancy of the raft was great enough to allow it to carry   lbs. over and above its own weight, which was ample to carry the machinery necessary to operate the dredge.

81

The striking features of this dredge are its durability and the ease with which it can be taken
apart and shipped to another territory. The rafts
of the great majority of the dredges built are
huge wooden hulls which cannot be used in more than
one place and are consequently a very heavy expense.

Bibliography.

The Metalurgy of Steel by Harbord and Hall.
The Manufacture of Iron & Steel Tubes by C.R.Works.
Kent's Mechanical Engineers Pocket Book.
Journa l of Am. Society of Naval Engineers, Nov.1911
National Tube Companies Book of Standards.
Trade Catalogs for the following:
    American Spiral Pipe Works,
    National Tube Co.,
    Stewarts & Lloyds,
    Mannesmann Tube Co.,
    East Jersey Pipe Co.,
    John Russel & Co.,
    Thomas Piggott & Co.,
    Abendroth & Root Co.,
    Standard Spiral Pipe Works,
    Robertson Bros. Mfg. Co.,
    American Welding Co.,
    Dayton Pipe Coupling Co.,
    S. R. Dresser Mfg. Co.,
The following Engineering Magazines.
    The Engineering Record,
    The Engineering News,
    The Electrical Age,
    The Canadian Engineer,
    The Iron Age,
    The Engineering Digest,
    The Boiler Maker.

WS - #0066 - 260225 - C0 - 229/152/10 - PB - 9781334182921 - Gloss Lamination